STRUCTURE OF *Matter*

TEACHER SUPPLEMENT

STRUCTURE OF *Matter*
TEACHER SUPPLEMENT

SCIENCE EDUCATION ESSENTIALS is a curriculum supplement series designed to cover vital topics in the various science disciplines, all from a thoroughly biblical viewpoint. Each product includes a teacher instructional guide along with K-12 activities and classroom helps to guide discussion, reinforce subject content, and facilitate hands-on laboratory exercises.

Published by the Institute for Creation Research.

SCIENCE EDUCATION ESSENTIALS

Series Creator:	Dr. Patricia L. Nason
Project Manager:	Janis McCombs
Managing Editor:	Beth Mull
Graphic Designer:	Susan Windsor
Science Reviewers:	Dr. John Morris, Dr. Charles McCombs, Dr. Randy Guliuzza, Dr. Larry Vardiman, Dr. Brad Forlow, Brian Thomas, Frank Sherwin

ISBN: 978-0-932766-96-0

STRUCTURE OF MATTER

Teacher Supplement Author:

Dr. Charles A. McCombs received his Ph.D. in Organic Chemistry from the University of California at Los Angeles in 1978. While at UCLA, Dr. McCombs studied new methods for the synthesis of bridged bicyclo-[2,2,2]-octenes via a novel Diels-Alder reaction in research, which led to the writing of two scientific publications on his work. He worked for 21 years as a Senior Research Chemist at Eastman Chemical Company Research group, where he specialized in vitamin E research and performed several special projects from which 20 patents were obtained.

In 1999, Dr. McCombs entered into Christian ministry, both as a creation speaker and Christian school administrator and teacher of math and science at the middle school, high school, and college levels. Dr. McCombs served as Research Associate at the Institute for Creation Research from 2008 to 2010.

K-12 Instructional Contributors: Dr. Patricia Nason, Dr. Daniel Criswell, Dr. Charles McCombs, Janis McCombs, Leona Criswell

For additional resources from the Institute for Creation Research, please visit www.icr.org or call 800.337.0375.

Copyright © 2010 by the Institute for Creation Research. All rights reserved. No portion of this book may be used in any form without written permission of the publisher, with the exception of brief excerpts in articles and reviews. For more information, write to Institute for Creation Research, P. O. Box 59029, Dallas, TX 75229.

ISBN: 978-1-935587-02-6

Printed in the United States of America.

TABLE OF CONTENTS

	PAGE
PREFACE	7
SCIENCE—A MATTER OF PREDICTION	9
FIRST AND SECOND LAWS OF THERMODYNAMICS	11
First Law of Thermodynamics	11
Second Law of Thermodynamics	13
STRUCTURE OF THE ATOM	17
Protons and Neutrons	18
Electrons	19
Quarks	21
THE PERIODIC TABLE OF ELEMENTS	25
Understanding the Elemental Box	26
A Survey of the Elements	27
Historical Development of the Periodic Table	29
The Problem of Isotopes	32
PROPERTIES OF CHEMICAL COMPOUNDS	35
Chemical Bonding	36
Reaction Rates	39
Phases of Matter	40
BIBLIOGRAPHY	43

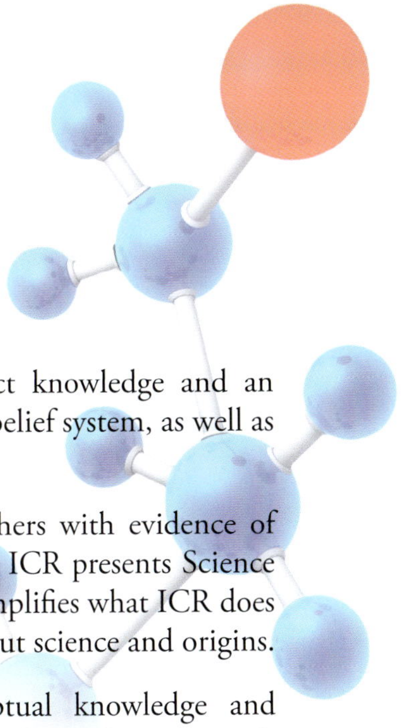

PREFACE

Teachers mold the minds of their students, helping them construct knowledge and an understanding of the world around them. A teacher's influence on the belief system, as well as cognitions, of a student can affect the student for a lifetime.

For 40 years, the Institute for Creation Research has equipped teachers with evidence of the accuracy and authority of Scripture. In keeping with this mission, ICR presents Science Education Essentials, a series of science teaching supplements that exemplifies what ICR does best—providing solid answers for the tough questions teachers face about science and origins.

This series promotes a biblical worldview by presenting conceptual knowledge and comprehension of the science that supports creation. The supplements help teachers approach the content and Bible with ease and with the authority needed to help their students build a defense for Genesis 1-11.

Each science teaching supplement includes:

- A content book written at the high school level to give teachers the background knowledge necessary to teach the concepts of scientific creationism with confidence. Each content book is written and reviewed by creation scientists, and can be purchased separately in class sets.

- A CD-ROM packed with teacher resources, including K-12 reproducible activities and PowerPoint presentations. The instructional materials have been pilot tested for ease in following instructions and completeness of activities. They have also been reviewed by scientists for scientific accuracy and by theologians for biblical correctness.

Science Education Essentials are designed to work within a school's existing curriculum, with an uncompromising foundation of creation-based science instruction. Secular textbooks are finding their way into Christian schools. Teachers may not lack belief in the Word of God, but they often do not have adequate information or knowledge concerning the tenets of scientific and/or biblical creation. Science Education Essentials equips teachers with the tools they need to teach the science of origins from a biblical rather than an evolutionary worldview.

The goal of each science supplement is to:

a) increase the teacher's understanding of and confidence in scientific creation and the truth of God's Word, while glorifying God as Creator;

   But sanctify the Lord God in your hearts: and be ready always to give an answer to every man that asketh you a reason of the hope that is in you with meekness and fear. (1 Peter 3:15)

STRUCTURE OF MATTER

b) provide teachers with a toolkit of activities and other instructional materials that build a foundation for their students in creation science apologetics;

Beware lest any man spoil you through philosophy and vain deceit, after the tradition of men, after the rudiments of the world, and not after Christ. (Colossians 2:8)

c) encourage the use of the higher level thinking necessary to stand firm against the lies of evolution and humanism.

…that we henceforth be no more children, tossed to and fro, and carried about with every wind of doctrine, by the sleight [trickery] of men, and cunning craftiness, whereby they lie in wait to deceive; but speaking the truth in love, may grow up into him in all things, which is the head, even Christ. (Ephesians 4:14-15)

With Science Education Essentials, teachers can equip the future generation of scientists and individuals to examine the evidence for the truth of Scriptures through an understanding of creation science. By using hands-on activities and relating scientific truth to the Bible, teachers/parents will be grounding their children in creation science truths so they can provide a logical response when challenged with science that is based on a philosophy that is in direct contradiction to Genesis 1-11.

As the leading creation science research organization, ICR is providing meaningful creation science material for classroom use. Our desire is that the materials renew minds, defend truth, and transform culture (Romans 12:1-2) for the glory of the Creator.

Dr. Patricia L. Nason

Science—A Matter of Prediction

Scientific predictions based on knowledge of observable events can be made with high probability. However, scientific predictions related to universal laws can be made with near certainty. Universal laws (e.g., the Law of Gravity, Laws of Thermodynamics) show highly predictable patterns of occurrence (behavior). This enables predictions of future events related to the unchanging principles of these laws to be made with near 100 percent accuracy.

The concept of the Law of Gravity—namely that what goes up must come down—is probably well understood. The Theory of Gravity became the Law of Gravity because all aspects of the theory were proven true at all locations and at all times. What would you think of the Law of Gravity if you shot an arrow into the air and it never returned to the ground? What would you think if you dropped an apple out of your hand and it flew up into the sky? What would you think if gravity changed according to the day of the week? These occurrences would show an inability to predict the effect of gravity. Predictability is the key to universal law. The certainty of scientific prediction related to the Law of Gravity is based on the universal, timeless predictability proven true in all locations and at all times.

There is certainty in scientific predictions related to universal law, as these processes are divinely ordered and orchestrated by God. As we will discuss throughout this material, the predicted certainty with regards to the structure and nature of matter reflects the created order established by God Himself at creation.

STRUCTURE OF MATTER

First and Second Laws of Thermodynamics

The Bible tells us in John 1 that Jesus is the Creator of all things. Everything in our universe came from only one source—our Creator God. If we believe that all "objects" on earth were made by God, then we must also believe that all "matter" on earth, the structure of that matter, and the laws of the universe governing that matter were also made by Him. As we study the structure of matter, we learn to recognize and appreciate the order and predictability of all of God's laws.

Matter: **a:** the substance of which a physical object is composed; **b:** material substance that occupies space, has mass, and is composed predominantly of atoms consisting of protons, neutrons, and electrons, that constitutes the observable universe, and that is interconvertible with energy.[1]

Two universal laws that are always in effect—and that are just as predictable as the Law of Gravity—are the First and Second Laws of Thermodynamics. Although affecting each of our daily lives, most people probably would not be able to provide an accurate description of the laws. In this lesson, we are going to learn more about these basic laws of physics that have been in effect since God created the heaven and the earth.

First Law of Thermodynamics

Sometimes it may seem to a child like there are a million rules that have to be followed. Do this and don't do that; stop this and stop that. When you were growing up, how many times were you told to turn off the lights, to close the door, or to turn off the water? Although rules at times are burdensome, we were all told these particular rules to conserve energy.

Conserving energy is important, but did you know that the conservation of energy is also a law of science? The First Law of Thermodynamics is a universal law that states that *matter and energy cannot be created or destroyed*. Matter and energy can exist in several different forms. Regardless of

First Law

Matter and energy cannot be created or destroyed.

---

1. Matter. *Merriam-Webster Online Dictionary*. 2010. Merriam-Webster Online. Posted on merriam-webster.com/dictionary/matter.

STRUCTURE OF MATTER

their forms, the total amount of matter and energy does not increase or decrease, but remains the same. For this reason, the First Law of Thermodynamics is also called the Law of Conservation of Energy.

This law has puzzled scientists for centuries. Scientists cannot explain how matter first appeared in the universe. In searching for this answer, little can be learned through observation since nothing is being created today. Equally puzzling to scientists is the fact that the total energy has not changed. The First Law of Thermodynamics states that there can be no change in the total amount of matter or energy—however, there is no law that tells us why. Science may tell us "how" something happens, but it can never tell us "why."

Where science fails, though, God's Word is always sufficient. When God created the heaven and the earth, He also created matter and energy. According to the First Law of Thermodynamics, all the matter

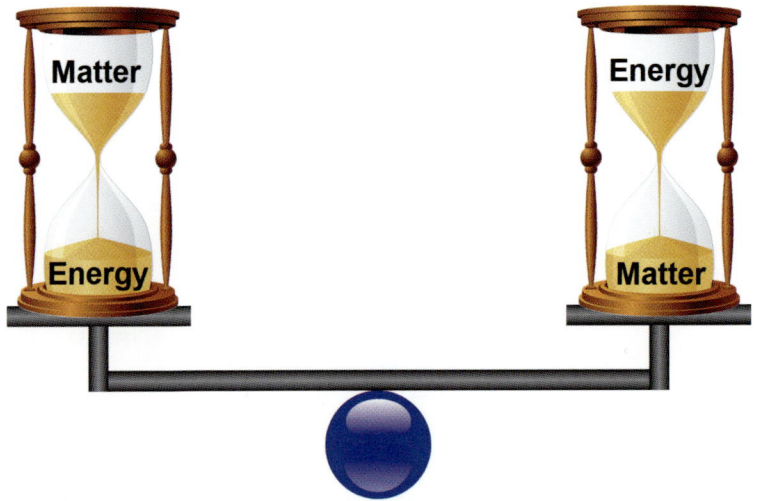

First Law of Thermodynamics: Matter and energy can interconvert, but the total amount does not change.

and energy that we will ever have was created in the beginning by God. Matter can be turned into energy and energy can be turned into matter, but no new matter or energy can be created.

The work of creation has been completed. Genesis 2:2 says that "God *ended his work* which he had made; and he rested on the seventh day from all his work which he had made." In doing so, God proclaimed that all creation of matter and energy had ceased. In accordance with God's Word, the First Law of Thermodynamics is being obeyed, as nothing new is being formed today by any processes.

Second Law of Thermodynamics

As a child, how many times were you asked to clean up your room? Guess what—every time your mother asked you this, she was trying to slow down the effects of the Second Law of Thermodynamics. This is a universal law of increasing *entropy*. Entropy is a scientific word for disorder or chaos. Therefore, the Second Law of Thermodynamics states that everything in the universe increases in disorderliness.

In other words, all things will constantly change or move in a "downward" direction toward being more disordered. All natural processes cause things to get worse with time, never better. Things will not "naturally" change from a state of chaos to a state of order. Change will only naturally occur from a state of order to one of chaos. In contrast, the theory of evolution describes a process in which all living things are constantly improving over time through a process of mutations and natural selection.

However, the only way to counteract entropy is to put extra energy into the process. In order for your room to become clean, energy must be added. Given the First Law of Thermodynamics, the energy required has to be produced through work. Consider the following statement by physicist Robert Lindsay, who speaks of the practical aspect of the Second Law in terms of "available" energy.

> It is in the transformation process that nature appears to exact a penalty and this is where the second principle [the Second Law] makes its appearance. For every naturally occurring transformation of energy is accompanied, somewhere, by a loss in the *availability* of energy for the future performance of work.[2]

Where does the work come from to produce a clean (ordered) room from disorder? Most likely from you, if you were being obedient to your

Second Law of Thermodynamics. Matter and energy lose order over time.

---
2. Lindsay, R. B. 1959. Entropy Consumption and Values in Physical Science. *American Scientist.* 47 (3): 378.

STRUCTURE OF MATTER

mother! However, where did you get the energy to produce the work? We receive energy from food. But where did the food come from? Can you see where these questions are leading? If you trace everything back to its origin, you will find that all current matter and energy came from the matter and energy created by God during the creation week.

> *"For by him were all things created, that are in heaven, and that are in earth, visible and invisible, whether they be thrones, or dominions, or principalities, or powers: all things were made by him, and for him: And he is before all things, and by him, all things consist."* Colossians 1:16-17

Although the total amount of matter and energy cannot be changed, some energy will be converted to heat. Heat is a form of energy, but heat cannot be converted back into work or into matter. According to Lindsay, for any work to be done, the "available energy" has to decrease because no process can be 100 percent efficient. Some of the available energy must be applied to overcome friction and some energy will be lost as non-recoverable heat.

Consider a car engine. As energy is applied and work is done, heat is generated but then lost in the exhaust. The generated heat cannot be reused to perform work once it is lost. Although the total amount of energy is conserved, the amount of available energy usable for work has decreased.

Everything in the physical universe is energy in some form. Every process loses "available" energy. Once processes continue long enough, *all* of the various forms of energy in the universe will have been converted to unavailable heat energy. At this point, no more work can be done and the universe will reach what the physicists call its ultimate "heat death."

The Second Law of Thermodynamics logically shows that the universe had a beginning. If there will be a day when the earth will "wind down," there must have also been a day when everything was "wound up"! We have already learned that the total quantity of energy in the universe is a constant, but the quantity of *available* energy is decreasing. The continued decrease in available energy implies that there was a time when all energy was available and when all energy was present in its original created form. This fact implies a beginning point—a day when all matter was created. Consistent with the Second Law of Thermodynamics, all of the available energy that was present in this world on that day began to become unavailable.

Everything that exists in the physical universe is in some form of energy. Everything that happens in the physical universe is a "process" that requires energy conversion. Therefore, the First and Second Laws of Thermodynamics that govern energy and energy conversion are the most fundamental of all scientific laws. Thus, the laws are extremely important for our understanding of the world around us.

Given these laws, how did the earth begin? Since the First Law of Thermodynamics is a universal law and matter and energy cannot be created or destroyed, it is only logical that the universe could not have begun by itself. Since energy cannot create itself and the energy required for creation must come from outside the universe, the most scientific and logical conclusion is that "in the beginning God created the heaven and the earth."

STRUCTURE OF MATTER

Structure of the Atom

In 1803, a Christian chemist named John Dalton proposed the atomic theory of matter. His discoveries are among the greatest achievements of modern chemistry and serve as the foundation for our understanding of matter. Dalton proposed that all matter consists of tiny particles called atoms. It is impossible to see an atom with the unaided human eye. Even the largest atoms known are sub-microscopic in size. Scientists know that atoms have a definite size and shape from studies using specialized equipment.

The word atom comes from the Greek word *atomos*. In Greek, the prefix *a* means "not" and the word *tomos* means "cut." Therefore, "atom" literally means "not-cuttable," or indivisible. An atom is considered indivisible because further cutting would divide its nucleus, changing the number of protons in the nucleus and subsequently the identity of the element of which that atom consisted.

To illustrate, consider a piece of copper wire. A copper wire is elementally just a collection of copper atoms attached to each other within a crystal structure. When a piece of copper wire is cut into two pieces, the result is just two smaller pieces of copper wire. Imagine cutting the end of the wire so that only one copper atom is de-

Structure of Matter

tached. Any further cutting would split that atom. We will learn later that the atom can be split into smaller pieces, but once an atom is split, its elemental identity has changed.

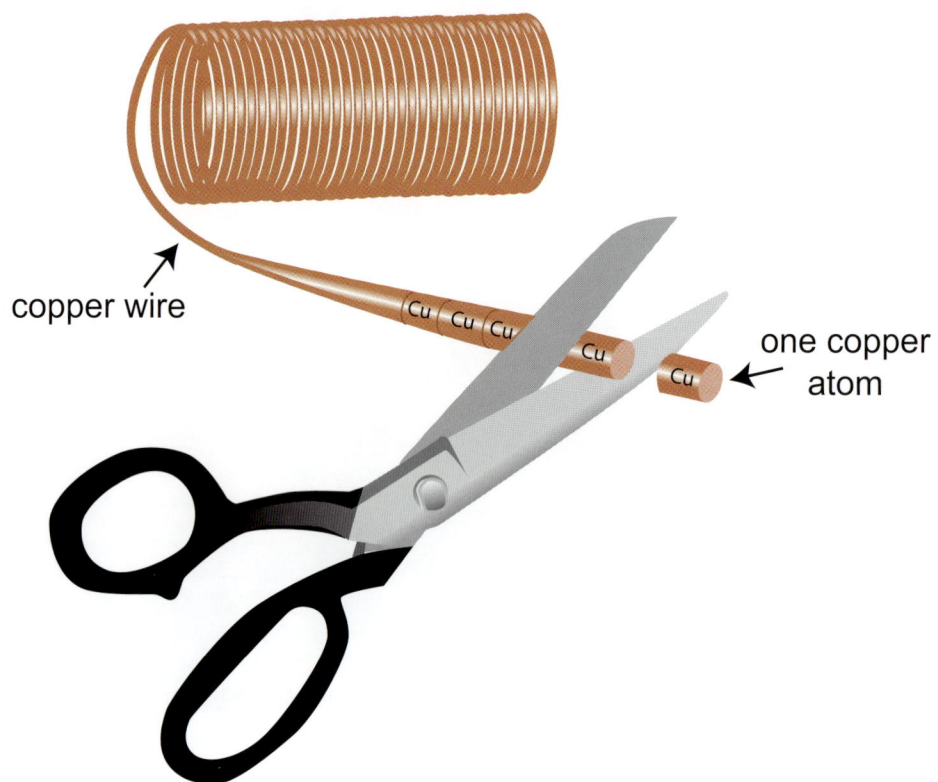

Protons and Neutrons

Dalton's atomic theory was a milestone in the study of chemistry. It gave scientists insight into how substances acquire their properties. However, his theory was not perfect. Dalton assumed that atoms were indivisible, like hard little spheres. In the late 19th and early 20th centuries, it became evident that atoms are composed of smaller particles of matter called subatomic particles.

The atom's dense central core, the nucleus, contains protons and neutrons. Protons carry a positive electrical charge and are probably the most important component of the nucleus. The number of protons determines the "identity" of an atom, and its identity determines the properties of that particular element. A hydrogen atom has a single proton; a helium atom, two protons; a lithium atom, three protons; and so on. A carbon atom is recognized as such because it has six protons in its nucleus. An atom is not carbon unless it has six protons.

The number of protons in the nucleus is referred to as the atom's *atomic number*. The atomic number is one of the important numbers on the periodic table of elements because each individual element has its own unique atomic number.

Neutrons have about the same mass as protons, but have no electrical charge and are therefore electrically neutral. Unlike protons, different numbers of neutrons can exist in atoms of the same element. However, the number of neutrons will never be fewer than the number of protons.

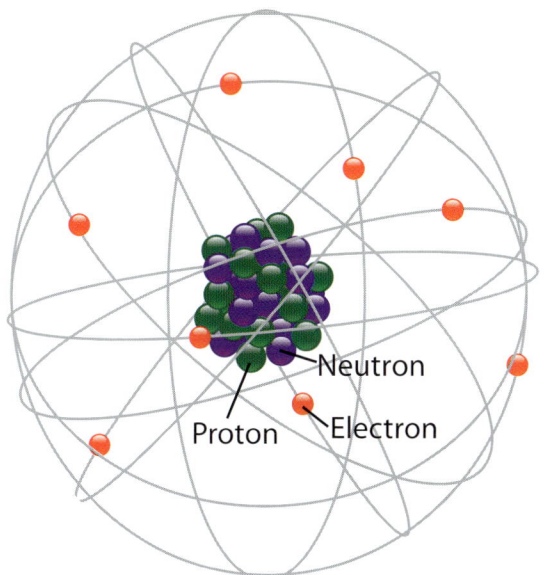

Structure of the atom (Note: Sizes and distance not to scale)

Electrons

The electron is the third component of the atom. Electrons carry a negative electrical charge equal in strength to the positive charge of a proton. For this reason, every uncharged element will have the same number of electrons as protons. Electrons are much smaller and lighter than protons and neutrons. The electron is actually 1,836 times less massive and thousands of times smaller than the proton.

Electrons always circle the nucleus. An atom is largely empty space because the electron orbits are located far from the tiny nucleus. If the nucleus were a golf ball at the center of a city, electrons would be a handful of sugar crystals blowing around the city in winds traveling near the speed of light.

The electron orbits the nucleus of an atom one hundred quadrillion (10^{17}) times per second. Although the electron at first seems to be moving in a random manner and in unpredictable orbits, electrons are actually arranged predictably in electron "shells" or energy levels that surround the nucleus. Each layer, or electron shell, can hold only a certain number of electrons. Electron shells farther from the nucleus can hold more electrons than the innermost shells. Electrons are not physically bound by the energy shell or energy level. As electrons absorb or release energy, they can change shells instantly.

Our understanding of electron motion is reflected in the quantum theory, which was developed in the early 1900s by Max Planck. The quantum theory states that tiny particles such as electrons do not absorb or release energy in a smooth or continuous manner, but energy is always absorbed or released in a discrete "packet" of energy called a *quantum* (plural is quanta). When an electron absorbs a quantum of energy, it moves farther away from the nucleus (higher energy level). When an electron loses a quantum of energy, it moves closer to the nucleus (lower energy level). In other words, electrons change shells, or energy levels, because they have either gained or lost energy.

Knowledge of the quantum theory led to the development of the atomic model. In 1913, the Danish physicist Niels Bohr (pronounced "Bore") proposed his model of the atom as a nucleus of protons and neutrons, surrounded by electrons that move in circular orbits at specific energy levels. The Bohr model gave scientists their understanding of atomic structure until the current model was developed with electrons residing in atomic orbitals, not circular orbits.

Orbital: a three-dimensional region surrounding the nucleus in which a particular electron can be found.

Figure 1 shows a depiction of the quantum model of a sodium atom. The atomic orbitals are described by quantum numbers, numbers that specify the properties of atomic orbitals and the properties of their electrons. The principal quantum number (also referred to as shells or orbitals) indicates the main energy levels surrounding a nucleus (represented by 1, 2, 3, 4, 5, 6, and 7). Larger numbers are farther away from the nucleus and electrons found here have more energy.

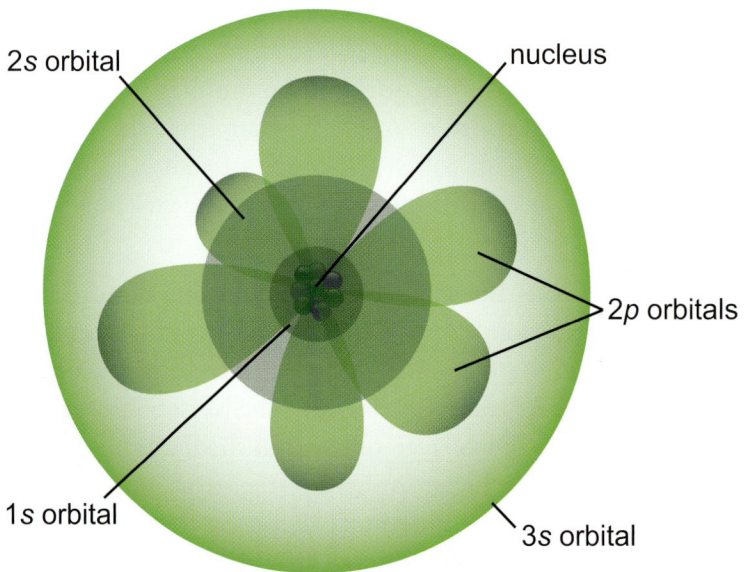

Figure 1: Quantum model of a sodium atom

The orbital quantum number (also referred to as sublevel or subshell) indicates the shape of the orbital (represented in order of increasing energy by s, p, d, or f). The *s*-orbital can hold two electrons, the *p*-orbital can hold six electrons, the *d*-orbital can hold 10 electrons, and the *f*-orbital can hold 14 electrons. The magnetic quantum number indicates the orientation of the orbital about the nucleus (represented by x, y, and z coordinates). The spin quantum number indicates the spin of an electron (represented by +½ or -½).

QUARKS

Earlier we learned that the word atom means not-cuttable, or indivisible. When we say that the atom is indivisible, we are saying that the atom cannot be divided into smaller pieces *and still retain the properties of the original element*. We now know that the atom contains subatomic particles in the nucleus called protons and neutrons. We are going to learn that the protons and neutrons themselves contain even smaller pieces of matter called quarks.

Do not be confused by the description of an atom as indivisible even though it contains smaller pieces. Although it is physically possible to break an atom into smaller pieces, the result is a loss of the atom's initial identity. For example, if we had one atom of gold, scientists today have the ability to break the gold atom into smaller pieces, but if they did, they would no longer have gold.

For many years, protons and neutrons were thought to be the smallest particles of the atom, although scientists suspected there were additional components. Scientists knew that protons have a +1 charge.

Fermi National Accelerator Laboratory (Fermilab)

STRUCTURE OF MATTER

Since each proton resembles a small magnet and magnets of the same charge always repel each other, there had to be a force holding the nucleus together that was strong enough to overcome the natural repulsive force of these magnets.

Almost since the discovery of the proton, scientists realized that there had to be some kind of "glue" holding the atom together. Scientists called this glue the *strong force*. Nuclear physicists reasoned that there must be something smaller than the proton or neutron in the nucleus. Using powerful particle accelerators (or "atom smashers"), scientists were able to detect particles even smaller than the proton and neutron.

In an atom smasher, particles are accelerated to speeds near the speed of light and then allowed to collide with a proton or neutron. If a proton or neutron contains smaller pieces, the atom smasher will break it into those pieces if enough energy is provided by the accelerated particle. Imagine a sealed paper bag with unknown contents being hit by a freight train at 100 miles per hour. Obviously, if the paper bag is empty, there would be no contents to see after the collision. However, if something is present in the paper bag, the contents would be seen after the collision. This is what happened in the particle accelerator. Scientists bombarded the nucleus with accelerated particles and the presence of smaller particles called quarks were detected.

Gluons (*glue* and the suffix *-on*) are elementary particles that are believed to be indirectly responsible for the binding of protons and neutrons in the nucleus. Physicists now believe the exchange of gluons between quarks in different protons and neutrons of the nucleus is the strong force (glue) that holds the nucleus together.

Although scientists have detected the presence of six quarks, there are two types of quarks that are important for this lesson: the "up" quark and the "down" quark. Both are about a third as massive as a proton or neutron. The up quark has an electric charge of +⅔, while the down quark has a charge of -⅓.

According to quark theory, it takes three quarks to make a proton or a neutron. A proton is composed of two up quarks and one down quark, giving the proton a combined electrical charge of +1. A neutron is composed of one up quark and two down quarks, giving the neutron a combined electrical charge of zero (electrically neutral). Individual quarks have never been observed or isolated. If quark theory is correct, quarks bind together so strongly that it is impossible to separate them from the other particles of the nucleus.

The quark theory is the predictable testimony of an orderly and purposeful Creator. The nucleus of the atom is a perfectly designed collection of particles that could never have been formed by chance or natural processes. The interactions between quarks and gluons hold the protons and neutrons together in the nucleus.

How did such a collection of particles come together to form a nucleus? To make the assumption that the nucleus was formed by natural processes presumes that one particle came first and the other particles then came together in a natural way. Protons challenge this premise. Whether first or last, the protons naturally would always get as far away from each other as possible because of their positive charges. There is no natural process to bring the protons into close proximity and have them remain there.

The only way for the nucleus of the atom to be formed is for all of the protons, neutrons, quarks, and gluons to come together at the same time. For one element to be formed in this manner, monstrous energy obstacles would have to be overcome. Each and every atom would have to go through the exact same uphill energy struggle, violating the First and Second Laws of Thermodynamics.

This method for building the nucleus could never be considered a natural process, because it cannot be explained by natural forces. Therefore, it must be considered a supernatural process. Science cannot explain this phenomenon using only natural laws. Science also cannot explain why the First and Second Laws even exist or how the first matter and energy came into being.

Evolutionary theory assumes that science must explain everything. However, science is the study of natural processes. Since protons would never get close together, neutrons have no reason to get close together because of their neutral charges, and the presence of quarks or gluons cannot change these facts, the logical conclusion is that the nucleus of each element was supernaturally created in its fully assembled form.

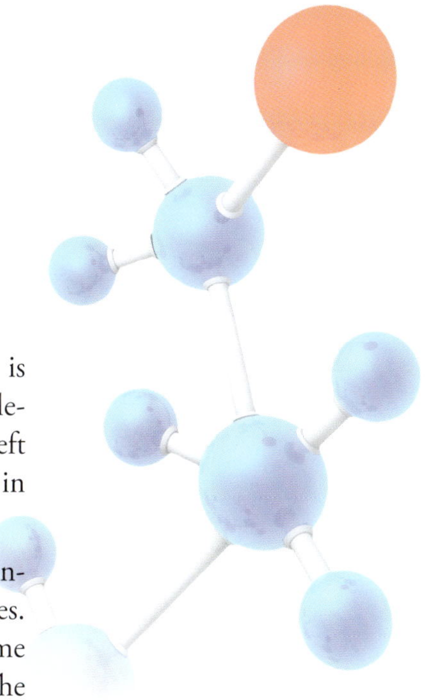

The Periodic Table of Elements

A table of the elements arranged by atomic number and valence is called the periodic table of elements. On the periodic table, the elements are arranged according to their atomic number in rows from left to right. By definition, the atomic number is the number of protons in the nucleus and is always represented by the letter Z.

The *periodic law* states that when elements are arranged by their increasing atomic numbers, they show repeating or periodic properties. Each horizontal row is called a *period.* Elements belonging to the same period have the same number of electron shells or energy levels. The first row is in the first energy level, the second row is in the second energy level, and so on.

Elements are also arranged vertically by *group.* Elements with the same group all have the same number of electrons in the outermost electron shell (valence electrons). The number of these electrons is called *valence.* The name of each element is given in its "box," along with its chemical symbol, atomic number, and average atomic mass.

- Nonmetals
- Alkali metals
- Alkaline earth metals
- Transition elements
- Other metals
- Metalloids
- Halogenes
- Noble gases
- Lanthanides
- Actinides

Period	Group 1 1A	2 2A	3 3B	4 4B	5 5B	6 6B	7 7B	8 8B	9 8B	10 8B	11 1B	12 2B	13 3A	14 4A	15 5A	16 6A	17 7A	18 8A
1	1 H 1.008																	2 He 4.003
2	3 Li 6.941	4 Be 9.012											5 B 10.81	6 C 12.01	7 N 14.01	8 O 16	9 F 19	10 Ne 20.18
3	11 Na 22.99	12 Mg 24.31											13 Al 26.98	14 Si 28.09	15 P 30.97	16 S 32.07	17 Cl 35.45	18 Ar 39.95
4	19 K 39.10	20 Ca 40.08	21 Sc 44.96	22 Ti 47.88	23 V 50.94	24 Cr 52	25 Mn 54.94	26 Fe 55.85	27 Co 58.47	28 Ni 58.69	29 Cu 63.55	30 Zn 65.39	31 Ga 69.72	32 Ge 72.59	33 As 74.92	34 Se 78.96	35 Br 79.9	36 Kr 83.8
5	37 Rb 85.47	38 Sr 87.62	39 Y 88.91	40 Zr 91.22	41 Nb 92.91	42 Mo 95.94	43 Tc (98)	44 Ru 101.1	45 Rh 102.9	46 Pd 106.4	47 Ag 107.9	48 Cd 112.4	49 In 114.8	50 Sn 118.7	51 Sb 121.8	52 Te 127.6	53 I 126.9	54 Xe 131.3
6	55 Cs 132.9	56 Ba 137.3	57 La 138.9	72 Hf 178.5	73 Ta 180.9	74 W 183.9	75 Re 186.2	76 Os 190.2	77 Ir 192.2	78 Pt 195.1	79 Au 197	80 Hg 200.5	81 Tl 204.4	82 Pb 207.2	83 Bi 209	84 Po (210)	85 At (210)	86 Rn (222)
7	87 Fr (223)	88 Ra (226)	89 Ac (227)	104 Rf (257)	105 Db (260)	106 Sg (263)	107 Bh (262)	108 Hs (265)	109 Mt (266)	110 Ds (271)	111 Rq (272)	112 Uub (285)	113 Uut (284)	114 Uuq (289)	115 Uup (288)	116 Uuh (292)	117 Uus 0	118 Uuo 0

6	58 Ce 140.1	59 Pr 140.9	60 Nd 144.2	61 Pm (147)	62 Sm 150.4	63 Eu 152	64 Gd 157.3	65 Tb 158.9	66 Dy 162.5	67 Ho 164.9	68 Er 167.3	69 Tm 168.9	70 Yb 173	71 Lu 175
7	90 Th 232	91 Pa (231)	92 U (238)	93 Np (237)	94 Pu (242)	95 Am (243)	96 Cm (247)	97 Bk (247)	98 Cf (249)	99 Es (254)	100 Fm (253)	101 Md (256)	102 No (254)	103 Lr (257)

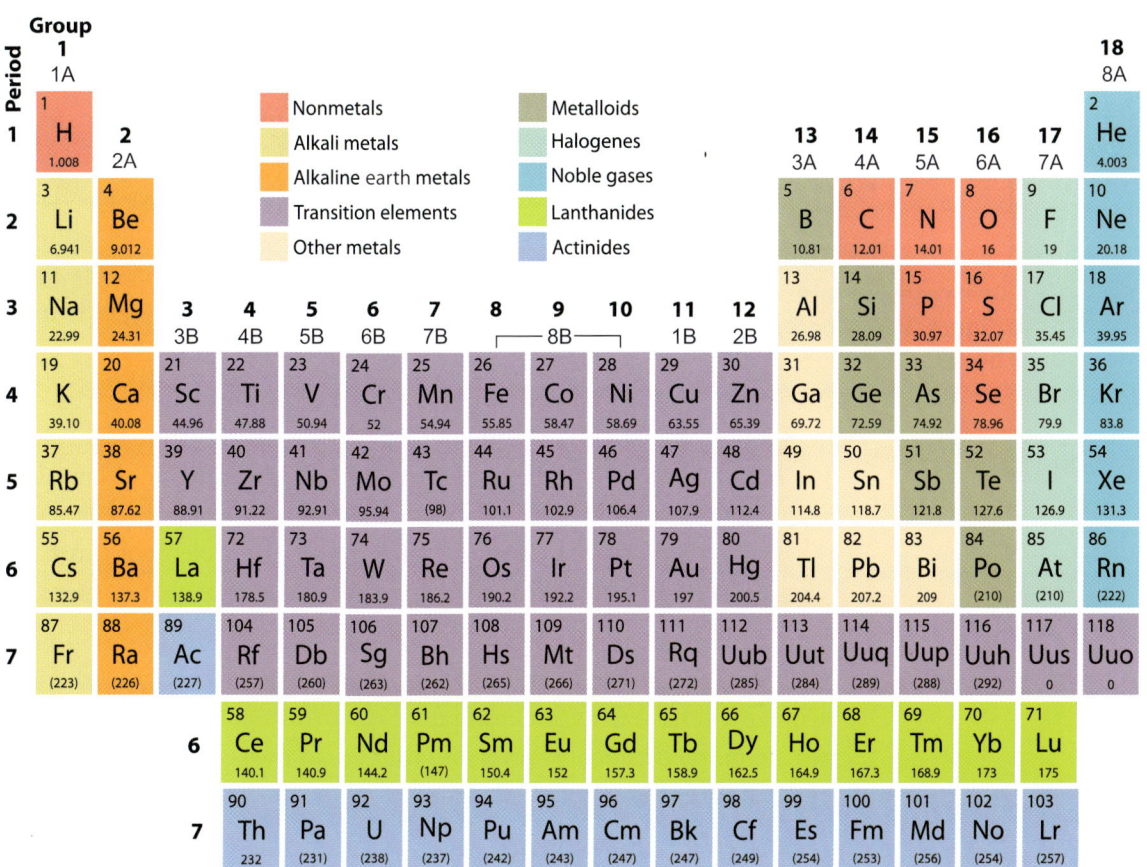

Understanding the Elemental Box

The atomic number for carbon is six. The atomic number of the element is usually located toward the top of the box or in a top corner (see Figure 2). Most periodic tables have a guide that shows the elemental information within the box. The periodic table is arranged numerically by atomic numbers. The decimal number, usually at the bottom of the elemental box, is called the *average atomic mass*. This number gives the average atomic mass of the element and its isotopes as they naturally occur. The decimal fraction results from a small contribution of atomic mass from other isotopes.

To understand the average atomic mass of an element, we must introduce two new terms called mass number and isotopes. The *mass number* of an element is the mathematical sum of the protons and the neutrons. Atoms of the same element that have different numbers of neutrons are called *isotopes*.

Figure 2: The carbon elemental box

For example, carbon atoms always have six protons, but may contain six, seven, or eight neutrons. Therefore, carbon has three isotopes: carbon-12, carbon-13, and carbon-14, with mass numbers of 12, 13, and 14, respectively. The most abundant form of carbon has six protons and six neutrons, giving a mass number of 12. The other isotopes of carbon cause the average atomic mass to be slightly higher than 12.

Naturally occurring carbon contains all three isotopes in the following abundances: 98.89 percent carbon-12, 1.11 percent carbon-13 and 0.00000000010 percent carbon-14. Based on the natural abundance of all three isotopes with the atomic mass of each isotope, the average atomic mass of carbon is calculated to be 12.01115 amu (atomic mass units).

Chlorine also exists as isotopes. Chlorine always has 17 protons, but may contain 18 or 20 neutrons. Therefore, chlorine has two major isotopes: chlorine-35 and chlorine-37. Many isotopes occur naturally, while others are man-made from other elements. Whether natural or man-made, each isotope of an element has basically the same chemical properties as other isotopes of that element.

Figure 2 indicates that carbon has six electrons. (The atomic number refers to the number of protons, but remember that every uncharged element will have the same number of electrons as protons.) Of the six, two are in the first energy level and four are in the second, outer energy level (i.e., are the valence electrons).

Groups are often numbered with the numerals one through eight. Groups in which the numeral is followed by the suffix A have the same

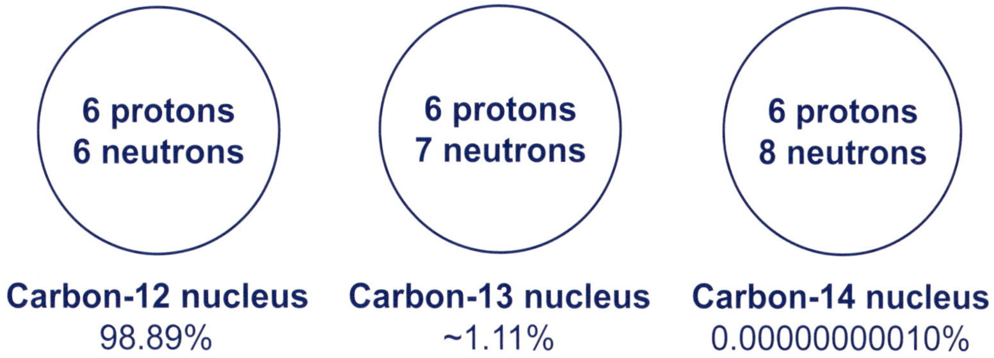

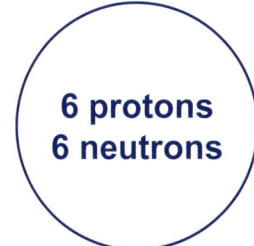
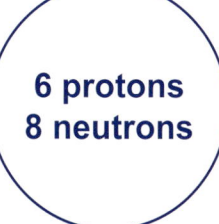

Carbon-12 nucleus   **Carbon-13 nucleus**   **Carbon-14 nucleus**
98.89%   ~1.11%   0.00000000010%

Relative Abundance of Naturally Occurring Carbon Isotopes

number of valence electrons as their group number. For example, elements in group 2A each have two outer shell electrons, while elements in group 7A have seven.

The elements in groups ending in the suffix B do not fit this pattern. These elements have either one or two valence electrons in their outermost shell. Elements in Group A have electrons filling *s-* and *p-*orbitals. Elements in Group B fill *d-*orbitals as well. The Group B elements of scandium through zinc have electrons filling the 3*d*-orbitals, but they all have one or two electrons filling each 4*s*-orbital because the 4*s*-orbital has lower energy than the 3*d*-orbital.

A Survey of the Elements

Elements with similar valences have similar properties. This feature helped early scientists predict unknown elements of the periodic table and helps chemists today predict the properties of an atom by comparing it to other atoms in the same group. There are eight groups of atoms, plus the transition elements.

Alkali metals. The elements of the periodic table are divided into several groups. The first group (Group 1A) consists of the *alkali metals.* These elements are very soft, shiny metals with low melting points (the temperature at which a solid turns into its liquid phase). All of these substances have only one electron in their outer shell. Many of these metals are important for our bodies in small amounts. For example, sodium and potassium help our bodies maintain a proper balance of fluids. The alkali metals react very easily with other atoms. As a result, they are corrosive and never appear naturally in their pure forms. Although hydrogen also has one electron in its outer shell, it is not included in the alkali metals because it has different properties.

Alkaline earth metals. Group 2A, the *alkaline earth metals,* all have two electrons in their outer shells. The alkaline earth metals do not react as readily with other substances as the alkali metals. Although harder and denser than the alkali metals, they are softer and less dense than most metals. All of these metals are silver-colored except beryllium, which is white. Beryllium and magnesium are prized for their strength and light weight, although beryllium is highly toxic. Powdered magnesium burns with a brilliant white flame and, as a result, magnesium is often used in signal flares and disposable photographic flashbulbs.

Transition metals. The *transition metals* include the groups ending in B (1B through 8B). All of these elements have either one or two electrons in their outer shells, and they all share similar properties. Most of these metals—such as titanium, iron, tungsten, and cobalt—are known for their strength and hardness. A few, such as copper and gold, are distinctively colored. Some transition metals are very durable and resistant to corrosion. Gold and chromium are often used to protect other substances from the elements. Because of their properties, the transition metals have a wide range of uses.

Inner transition metals. Beneath the main body of the periodic table lies a group of 30 metals called the *inner transition metals*. These metals actually fit between Groups 3B and 4B. The inner transition metals have two electrons in their outer shells and share similar properties. Many of these elements are radioactive. One of these radioactive elements, americium, is used in household smoke detectors. Other well-known radioactive substances are plutonium and certain isotopes of uranium, both of which are widely used in nuclear reactors.

Mixed groups. Groups 3A, 4A, and 5A do not have special names like the other groups. Composed of nonmetals, semimetals, and metals, these groups exhibit a wider variety of properties than the other groups. Group 6A elements are sometimes called *chalcogens* or the *oxygen family*.

These groups include some of the most important elements to man. Oxygen (Group 6A) is vital to animal and human life. Carbon (Group 4A) is a basic constituent of our bodies and is also found in most fuels (such as petroleum and coal). One form of carbon, called *graphite*, is a soft, crumbly black solid used as a powdered lubricant and as pencil "lead."

Carbon also appears in a transparent crystalline form called *diamond*, the hardest substance known to man and one of the most beautiful. Diamonds are widely used in industry, as well as in fine jewelry. Silicon (Group 4A), found in beach sand, is commonly used in glass, computer chips, and solar cells.

Elements in Group 7A are known as the *halogens*, meaning "salt formers." They easily combine with Group 1A and 2A metals to form salts. Halogens are very useful. Fluorine is added to toothpaste and drinking water because it hardens tooth enamel, hindering tooth decay. Chlorine is often used in the purification of water to kill bacteria and other disease-causing organisms and to remove bad tastes and odors. Although very corrosive to the skin, bromine is used to kill fungus infections in plants. Iodine is commonly used for disinfecting and is also important in the human diet. Halogens are also used in certain types of incandescent light bulbs (halogen bulbs) to prolong the life of the filament.

Noble gases. The noble gases (Group 8A) rarely combine with other elements because they already have eight electrons in their outer shells (two electrons in helium). These elements have very similar properties. All are gases, and all are found in nature in their pure states. No natural compounds of the noble gases exist.

The fact that they do not combine easily with other substances makes these elements very useful. Incandescent lighting with argon or krypton is preferred over air-filled lighting because they do not break down the filament at high temperatures. For the same reason, certain materials are bathed in argon gas while being welded.

Helium, which is lighter than air, is used to fill airships and helium balloons and as a low-temperature coolant. Various noble gases are also used in neon signs, with different gases giving different colors. Airport runway lights use noble gases as well.

The first 92 elements of the periodic table represent all of the naturally occurring elements known to man. The elements from 93 through 118 (total as of 2008) were formed by fast atom bombardment and hence have only been observed in the laboratory.

Today, scientists have a good understanding of the periodic table of elements, but that knowledge did not come easily. Although German alchemist Hennig Brand first discovered phosphorus in 1669, it would be another 200 years before Russian chemist Dmitri Mendeleev would be given credit for creating the first version of the periodic table of elements.

HISTORICAL DEVELOPMENT OF THE PERIODIC TABLE

During the 17th century, there were very few chemical reactions known and most laboratory work was trial and error. Scientists knew of gold, silver, tin, lead, copper, and mercury, but they did not associate these metals with being chemical elements. When Hennig Brand tried to turn base metals into gold (a practice known as alchemy), the hot residues obtained by boiling urine formed a liquid that burst into flames.

That liquid was elemental phosphorus. Since phosphorus was made by a chemical reaction, scientists began to study chemistry in a different way, learning how to make other elements. In the 17th century, chemists only had around 16 known chemicals to work with in the laboratory, and not all of these were elements. By 1806, chemists had discovered 47 elements. As more and more elements became known, chemists began to notice that several elements had similar properties, and they placed these elements into groups.

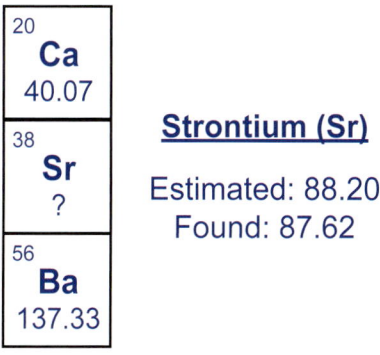

Alkaline earth triad

STRUCTURE OF MATTER

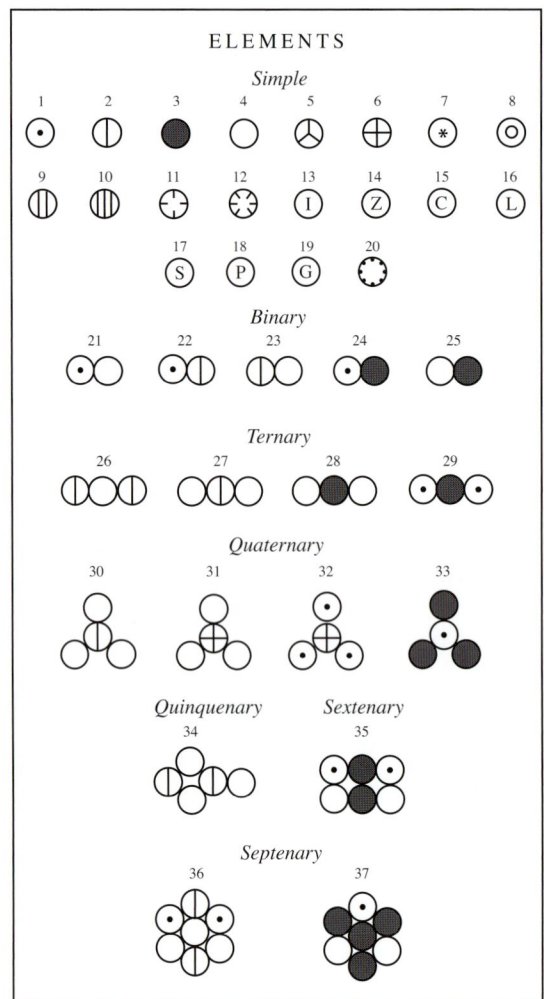

The elements and molecules depicted by English chemist John Dalton[3]

In 1817, German chemist Johann Dobereiner noticed, in one of these groups, that the atomic weight of strontium fell midway between the atomic weights of calcium and barium, and that all three elements of this triad possessed the same chemical properties. Later, the halogen triad of chlorine, bromine, and iodine, and the alkali metal triad of lithium, sodium, and potassium were observed. From these triads, Dobereiner proposed the Law of Triads, which held that the middle element of a triad had properties that were an average of the atomic weights of the other two members.

Between 1829 and 1858, the study of triads became increasingly popular and scientists soon realized that the similarity of elemental properties extended beyond just groups of three. Fluorine was added to the halogen group of chlorine, bromine, and iodine. Oxygen, sulfur, selenium, and tellurium were grouped into a family, and nitrogen, phosphorus, arsenic, antimony, and bismuth were classified as another.

---

3. Image adapted from page 227 the 1842 reprint of Dalton's 1808 book *A New System of Chemical Philosophy*. He wrote, "This plate contains the arbitrary marks or signs chosen to represent the several chemical elements or ultimate particles." Mendeleev's 1869 chart was more complete and replaced the arbitrary symbols with the letters still in use in today's periodic table of elements.

Series	Group I	Group II	Group III	Group IV	Group V	Group VI	Group VII	Group VIII
1	H=1							
2	Li=7	Be=9.4	B=11	C=12	N=14	O=16	F=19	
3	Na=23	Mg=24	Al=27.3	Si=28	P=31	S=32	Cl=35.5	
4	K=39	Ca=40	— =44	Ti=48	V=51	Cr=52	Mn=55	Fe=56, Co=59, Ni=59, Cu=63.
5	(Cu=63)	Zn=65	— = 68	— = 72	As=75	Se=78	Br=80	
6	Rb=85	Sr=87	?Yt=88	Zr=90	Nb=94	Mo=96	— =100	Ru=104, Rb=104, Pd=106, Ag=108.
7	(Ag=108)	Cd=112	In=113	Sn=118	Sb=122	Te= 125	J=127	
8	Cs=133	Ba=137	?Di=138	?Co=140	—	—	—	— — — —
9	(—)	—	—	—	—	—	—	
10	—	—	?Er=178	?La=180	Ta=182	W=184	—	Os=195, Ir=197, Pt=198, Au=199.
11	(Au=199)	Hg=200	Tl=204	Pb=207	Bi=208	—	—	
12	—	—	—	Th=231	—	U=240	—	— — — —

Mendeleev's 1869 periodic table of elements

In 1863, English chemist John Newlands wrote a paper that classified the 56 known elements into 11 groups based on similar physical properties. In his studies, he observed that elements with similar properties existed that differed by multiples of eight in atomic weight. In 1864, Newlands published his version of the periodic table and proposed the Law of Octaves: Any given element will exhibit analogous behavior to the eighth element following it in the table.

The Law of Triads clearly showed that there were several vertical patterns of predictability within the known elements. The Law of Octaves showed for the first time that there was also a horizontal pattern within the known elements. Today, it is known that there are diagonal trends within the table.

Recognition of these patterns created an excitement within the scientific community. Many scientists were working competitively to produce a working version of the periodic table. However, Mendeleev's version published in 1869 became the model for the periodic table that we know and use today. Although Mendeleev was given credit for the discovery, his table did not look anything like what we have today. To Mendeleev's credit, his insight gave chemists the ability to predict that elements with certain atomic weights and respective properties would be found. Mendeleev's table demonstrated the periodic nature of the elements, and discoveries by 20th-century scientists explain why the properties of the elements recur periodically.

In 1911, English chemist/physicist Ernest Rutherford and German physicist Hans Geiger discovered that electrons orbit the nucleus of an atom. In 1913, Niels Bohr discovered that electrons move around the nucleus in discrete energy levels. Electrons do not circle the nucleus in orbits like the moon around the earth, but move about freely within regions called orbitals.

Rutherford first identified protons in the atomic nucleus in 1914. These were important discoveries because until that time, scientists studied the atoms by their properties, not by their structure. In 1932, English physicist James Chadwick first discovered neutrons, and isotopes were identified.

With the discovery of electrons, protons, neutrons, and isotopes, scientists now had the complete basis for the periodic table. With the discovery of isotopes, scientists began to understand why certain trends did not fit their predictions. Their predictions were based on atomic mass, which can be heavily influenced by the presence or lack of isotopes. It quickly became apparent that the properties of the elements varied periodically with atomic number, not the atomic mass.

The last major change to the periodic table resulted from American chemist Glenn Seaborg's work in the mid-20th century. Starting with plutonium in 1940, he discovered all the transuranic elements from 94 to 102. In 1945, Seaborg identified lanthanides and actinides ($Z > 92$), and placed the actinide series below the lanthanide series. In 1951, he was awarded the Nobel Prize in Chemistry. For his work, element 106 was named seaborgium (Sg) in his honor.

The Problem of Isotopes

As early scientists were developing the periodic table, they tried to show a periodicity based on the atomic mass of the element. The atomic mass is essentially the combined mass of the protons and neutrons in the nucleus. If they grouped the elements by properties, the atomic masses were not sequential. If they sequenced the elements by atomic mass, not all elements within a vertical group had the same properties.

This caused early scientists a problem in finding a periodicity with atomic mass. What caused this problem? The answer is isotopes. Isotopes contain extra neutrons, resulting in increased atomic mass. The actual atomic mass is the weight of the protons and all of the neutrons present in the nucleus in their naturally occurring proportions. Because they contain isotopes, hydrogen has an atomic mass of 1.008 instead of 1.000 and carbon has an atomic mass of 12.01 instead of 12.00.

Note that the first ten elements of the periodic table appear to have equal numbers of protons and neutrons. The atomic mass is almost exactly twice the atomic number (except for hydrogen). Let's compare two heavier elements. Argon ($Z = 18$) has an atomic mass of 39.95. But the next element, potassium ($Z = 19$), has an atomic mass of 39.098. The element with the smaller atomic number has a larger atomic mass.

Based on the factor of two times the atomic number, potassium would be expected to have an atomic mass of 38.000, but there is 1.098 extra mass. Using this same factor, argon would have a theoretical atomic mass of 36.00, but argon has 1.95 extra mass. Uranium (Z = 92) would have an atomic mass of 184.00, but its atomic mass is actually 238, an extra mass of 54. The extra mass is the mass of the extra neutrons. Scientists now define "relative atomic mass" as the average mass of all the isotopes present for that element.

The periodic table of elements is truly a superb example of God's amazing creation. All elements have the same components (protons, neutrons, and electrons), but each element is different from the next. Each component of the atom is simple in structure, but we still do not know everything about them.

The periodic table would appear to be simply a numerical sequence or a list of elements, but the chemical properties of many unknown elements were predicted before they were discovered. The periodic table is complex, but easily understood. This unique collection of atoms—whose identity is determined solely by the number of protons in the nucleus and the predictability of properties found in these atoms—could never be the product of random chance. This order and predictability could not have just happened.

Remember, the difference between any two adjacent elements in the periodic table is just one proton. Think about the difference of only one proton between the toxic gas chlorine and the inert gas argon. The orderly sequence of the elements is a reflection of an orderly and systematic Creator. Without the predictability found in the periodic table, we still might be searching for the elements. The highly predictable arrangements of the elements by period and group within this table reveals the order of an omnipotent Creator God.

Properties of Chemical Compounds

John Dalton's atomic theory explained many of the mysteries of chemistry by demonstrating that atoms can be combined in various ways to form different substances. Elements are composed of a single type of atom. Compounds are composed of more than one type of atom linked together. Elements, compounds, and molecules are the building blocks of every chemical substance known to man. The foods we eat, the clothes we wear, and the things we use (e.g., camera, computer, medicine, or books) are all examples of chemical compounds.

Gold and silver metal are two uncombined elements found naturally in their pure form. Metals like aluminum and nickel are found as compounds and need to be chemically converted to produce useful metals. Whether found in a natural form or chemically converted, all of the elements play an important part in chemistry. Chemical processes can be explained as the rearrangement, combination, or separation of atoms. According to the atomic theory, the differences between various kinds of matter result from differences in their respective atoms or in the way their atoms are combined into molecules.

Nuclear power, nuclear medicine, lasers, and numerous other discoveries have been made possible by our understanding of atomic structure. Chemists know *how* substances will behave because of the predictable nature of the atom. The *properties,* or characteristics, of an element depend largely on the number of electrons contained in its outer electron shell. Even though some electron shells could theoretically hold more than eight electrons, the last electron shell of an atom never holds more than eight electrons. The electrons in the outermost shell are known as the valence electrons. The number of these electrons, or *valence,* largely determines an element's properties.

Since every element of the periodic table possesses protons, neutrons, and electrons, the properties of the different elements can only be determined by the number and/or arrangement of the subatomic particles. The number of protons gives the scientist the identity of the element, but it is the number of electrons in the outer shell (valence) that determines how each atom reacts and behaves.

Atoms may gain or lose their outermost electrons through a variety of processes such as heat, electricity, radiation, or chemical action. An atom that loses electrons will no longer balance; it will have more pro-

STRUCTURE OF MATTER

tons than electrons, and thus possess a positive charge. Likewise, an atom that gains electrons will have more electrons than protons, resulting in a negative charge. An atom that develops an electrical charge by losing or gaining electrons is called an *ion*. Negative ions are called *anions* and positive ions are called *cations*. When a substance consists of ions, it is said to be ionized.

Regardless of how an atom reacts or what changes are made by a reaction, *atoms generally seek to have eight electrons in their outermost shell.* Atoms that already have eight electrons in their outer shells tend not to combine with other atoms. Atoms smaller than boron (Z = 5) cannot have eight electrons in the outer shell. They would rather donate their outer electrons, leaving their next lower electron shell full (two electrons).

Some atoms can gain electrons and some atoms can lose electrons, but a chemical reaction will only occur if both atoms end up with a total of eight electrons in their outer shells (two electrons for the small atoms). This simple fact is the basis of all modern chemical reactions. In order to have a better understanding of chemical reactions, we need to understand chemical bonding and reactions rates.

Chemical Bonding

A *chemical bond* is the force of attraction between atoms. Chemical bonding occurs when atoms either transfer or share electrons in order to have eight electrons in the outer shell. All chemical bonds known today can be classified as either ionic or covalent bonds. Bonds formed by the transfer of electrons are called *ionic bonds*. Bonds formed by the sharing of electrons are called *covalent bonds*.

Ionic bonds are formed by the transfer of electrons from one element to another. Typically, ionic bonds are formed between elements from Group 1, 2, or 3 with elements from Group 6 or 7. The atom that loses electrons will have a positive charge and the atom that gains electrons will have a negative charge. Therefore, the two atoms, now with opposite charges, attract each other like poles of a magnet. Common household examples of ionic compounds include lithium fluoride (used in toothpaste), sodium bicarbonate and carbonate (used as baking powder and baking soda), magnesium sulfate (used in foot soaking baths), and sodium chloride (table salt).

Ionic bond: a chemical bond formed by the transfer of electrons from one element to another. **Covalent bond:** a chemical bond formed by the sharing of electrons from one element to another.

Sodium and chlorine atoms do not contain eight electrons in the outer valence shell of electrons in their elemental state (Figure 3). They are,

ELEMENTAL STRUCTURE (UNSTABLE)

SODIUM ATOM
Na
(11 Electrons)

CHLORINE ATOM
Cl
(17 Electrons)

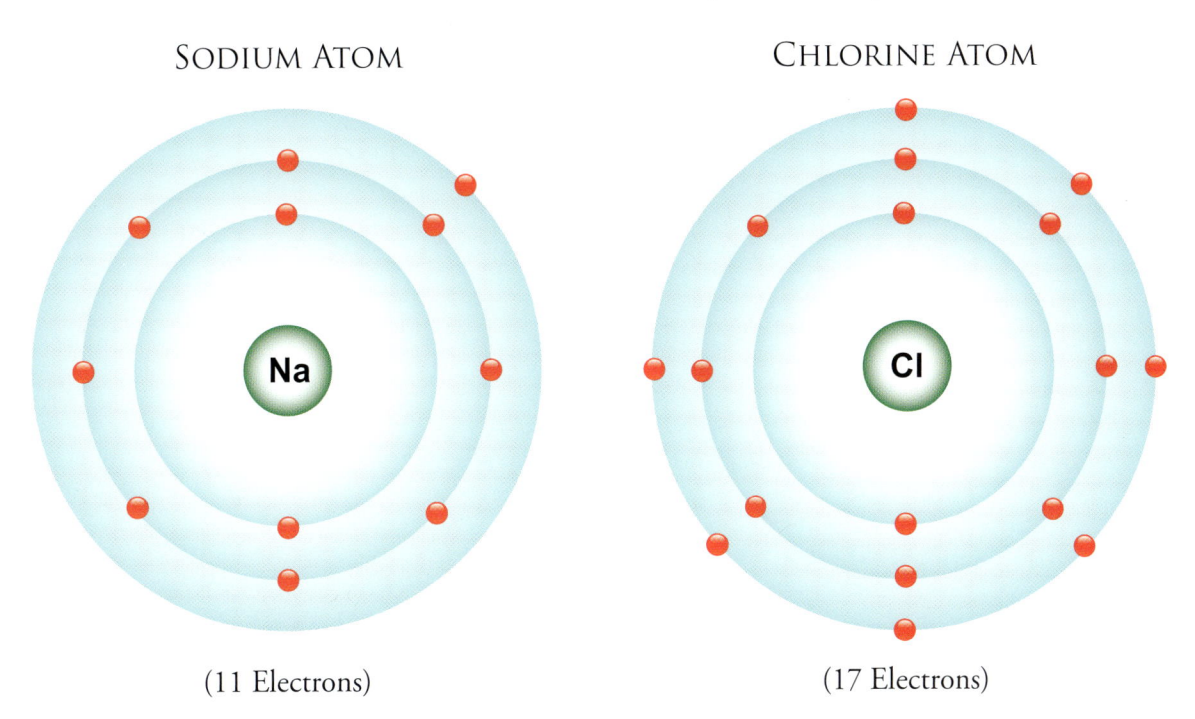

Figure 3: Bohr models of sodium and chlorine

therefore, very reactive elements and will react in any manner that will provide eight electrons in the outer shell. Consequently, sodium is violently reactive in the presence of water (giving its electron to water). Chlorine is the active ingredient of bleach and a powerful disinfectant (pulling an electron away from something else).

Consider the reaction between sodium and chlorine. The transfer of one electron from sodium to chlorine changes the valence (and properties) of both atoms. Sodium (a Group 1A element) is willing to donate one electron in order to end up with eight valence electrons. Chlorine (a Group 7A element) is willing to accept one electron, resulting in eight valence electrons. As a compound, each atom now has eight electrons in its outer electron shell. Consequently, both are stable and no longer reactive like the natural elements.

When a sodium atom donates one electron, it becomes a sodium ion. Likewise, when a chlorine atom accepts one electron, it becomes a chloride ion. Figure 4 shows that the sodium ion no longer has any electrons in its original outer shell. The one it originally had in the outer shell is the one that was donated to chlorine. Therefore, the sodium ion now has a full (eight electrons) outer shell. By accepting the electron from sodium, the chlorine ion also has a full outer shell. The newly formed sodium ion and chloride ion both have eight electrons in their outermost valence shell of electrons, just not in the same shell. The resulting ionic structure is called sodium chloride.

Ionic compounds such as sodium chloride and other salts are compounds formed with ionic bonds. Elements from opposite ends of the periodic

Ionic Structure (Stable)

Sodium Ion

Na

(10 Electrons)

Chloride Ion

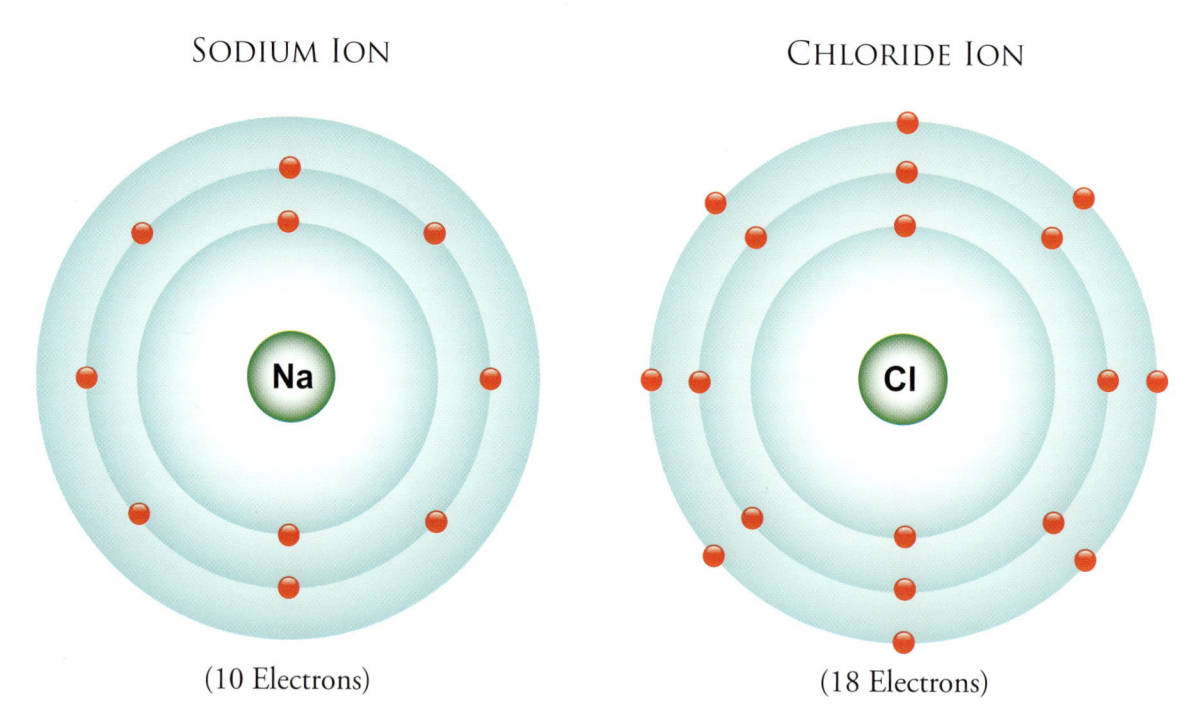

Cl

(18 Electrons)

Figure 4: Sodium chloride

table tend to form ionic compounds, which tend to be brittle, often have a high melting point, and are in a solid state at room temperature. Many ionic compounds can be dissolved in water and the resulting solution can conduct electricity. The atoms in an ionic compound are not really individual molecules, but a three-dimensional stacking of the individual ions into what is called a *crystal structure*.

In a covalent bond, electrons are shared between the two atoms, not owned by one particular element. Common household examples containing covalent bonds include vinegar, butter, cooking oil, wax, and water. The structure of a covalent compound is very different from that of an ionic compound.

The transfer of one or two electrons is easy to do electronically. However, the transfer of two or more electrons requires much more energy. In this case, atoms often share the electrons instead of transferring them. Water, which contains the elements hydrogen and oxygen, is an example of a molecule formed with covalent bonds. In the formation of water from hydrogen and oxygen, the electrons are equally shared and do not belong to either hydrogen or oxygen. An oxygen atom (in Group 6A) normally contains six electrons in its outermost shell, with two unpaired electrons available for bonding (Figure 5). Elemental hydrogen has one unpaired electron. Therefore, when two hydrogen atoms and one oxygen atom react, they share their electrons and fill their outermost electron shells.

Covalent compounds often have a lower melting point than ionic compounds and usually do not conduct electricity. Many covalent com-

38   STRUCTURE OF MATTER

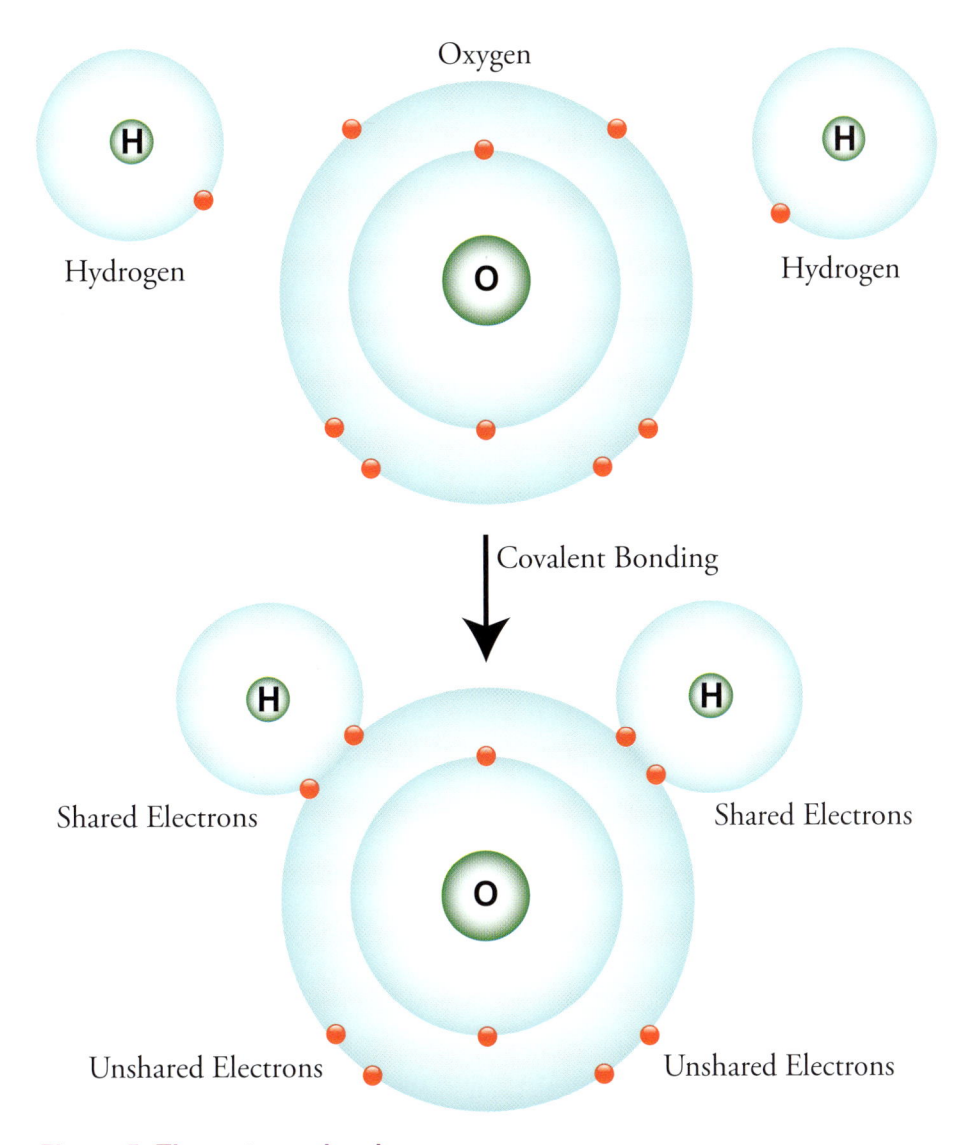

Figure 5: The water molecule

pounds are not water soluble, although some of the smaller or more polar molecules like sugar are. The closer two non-metals are together within the periodic table, the more likely those bonds will form a covalent compound.

Reaction Rates

Kinetics is the study of reaction rates and provides a very powerful tool for understanding how certain elements and compounds respond in chemical reactions. All molecules in a gas or liquid phase have atoms that move freely in their environment. The atoms of solid compounds are bound in a crystal structure. All atoms and molecules have the possibility of colliding with other atoms or molecules. If colliding atoms have no mechanism for electron transfer or sharing, they will only bounce off each other, much like the collision of pool balls on a pool table.

Nitrogen and oxygen are two gases contained in our atmosphere that frequently collide with each other. Because there is no mechanism for

STRUCTURE OF MATTER

interaction, these molecules never react with each other. However, if there is a mechanism for the individual atoms to acquire eight valence electrons during a collision, a chemical reaction will take place. The atoms will either transfer or share the electrons, depending on the identity of the atom and the properties of that element.

When a chemical reaction is the result of a collision between two molecules, the reaction rate is the rate at which the collision occurs. The reaction rate can be increased by changing the concentration, temperature, or pressure of the system. An increase in the concentration of a system (by decreasing the volume or by increasing the pressure) effectively makes a more crowded environment, increasing the likelihood of a collision.

Heat, radiant energy, or anything that increases the temperature also causes an increase in the reaction rate because faster-moving molecules are able to find other molecules more readily. In general, anything that can be done to increase the probability of two molecules colliding with each other will increase the reaction rate of the chemical reaction.

The shapes of chemical molecules, especially large molecules, can also play an important part in chemical reactions and chemical properties. The size of large molecules makes it possible that the shape will affect the molecule's ability to interact with other molecules. Sometimes the shape can play a positive role, but sometimes the shape has an adverse effect.

A thorough knowledge of kinetics and reaction rates is an important tool because it allows the chemist to predict and control chemical reactions. All chemical reactions are governed by reaction rates. Chemical reactions do not just happen by chance. They occur according to known chemical principles, to predictable properties and established reaction rates. Knowledge of chemistry gives us the ability to understand more about our Creator who *created all things*, including matter and energy, and *in whom all things consist*. Without him, we could not exist!

Phases of Matter

Matter can exist in various distinct forms, or phases (also referred to as states). Examples of extreme phases include plasma, critical fluids, and degenerate gases. However, the three phases that substances commonly exist in are solid, liquid, and gas.

In solids, the molecules are tightly packed together in essentially fixed positions (cannot move around), resulting in an ordered arrangement (Figure 6). Examples of solids include rocks, wood, salt, and ice.

In liquids, the molecules are no longer in a fixed position, resulting in a more disordered structure. Although the molecules are still close to each other, they are free to slowly move around relative to each other. Examples of liquids are mercury (at room temperature), gasoline, and water.

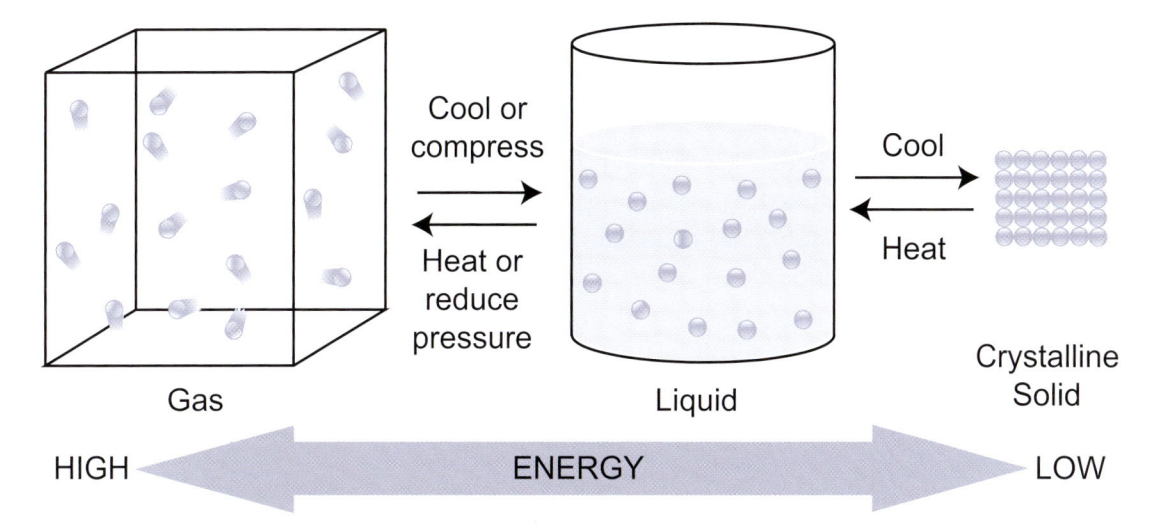

Figure 6: Phases of matter

In gases, molecules are widely separated with much empty space between them, resulting in total disorder. Therefore, the molecules have complete freedom of motion and move around at high speeds. Examples of gases are nitrogen, oxygen, helium, and steam.

The phase of a substance depends on the potential energy in the atomic forces holding the molecules together and on the thermal energy of the molecule motions. Therefore, the phase of matter is dependent on the temperature and pressure. A phase diagram of a substance shows the phase (solid, liquid, or gas) it exists in at a given temperature and pressure. Figure 7 shows a phase diagram of water.

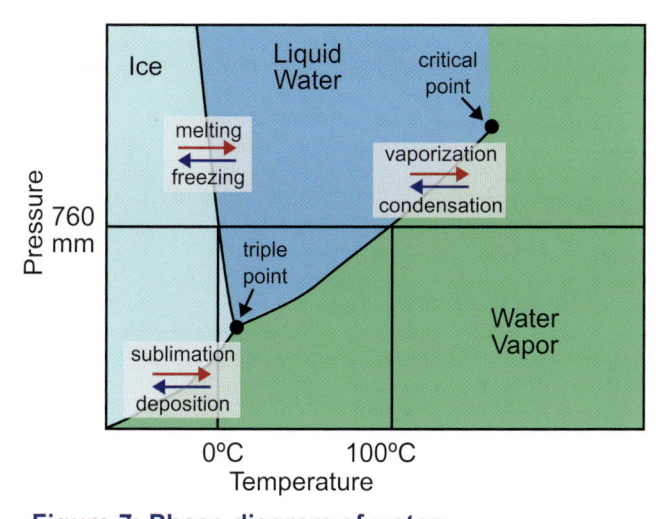

Figure 7: Phase diagram of water

The three lines constructing the phase diagram identify the interfaces between two phases—solid/liquid, liquid/gas, or solid/gas. At these corresponding temperatures and pressures, both phases exist in equilibrium. The triple point is the temperature and pressure at which the three lines meet. At this temperature and pressure, all three phases of matter can exist. The critical point identifies the temperature and pres-

STRUCTURE OF MATTER

sure at which the liquid and gaseous phases of a pure, stable substance become identical.

As the temperature and pressure conditions change, matter may change from one phase into another. This change in structure and properties is referred to as a phase transition. On the phase diagram, the phase transitions are depicted using red and blue arrows. For example, as an ice cube (solid water) is heated at standard atmospheric pressure (760 mm Hg), the ice will melt into liquid water once the temperature is greater than 0°C. If the liquid water is heated to a temperature greater than 100°C, the water vaporizes into steam.

As you can see, the structure of matter is not simple. Its design is complex, yet the properties can be predicted. We do not need a microscope or laboratory to see God's omnipotent power and omniscient strength. Open your eyes and look around. God's fingerprints are all over His creation. The evidence is not hidden, but rather in clear sight for everyone to see.

Although some scientists choose to believe otherwise, Romans 1 says that they are "without excuse" and have "changed the truth of God into a lie, and worshipped and served the creature more than the Creator." When you look at the clear evidence around us, you will see that chemistry and the structure of matter are truly God's testimony to man. God's Word is true and accurate. The structure of matter and the Bible are both proclaiming the same truth—that "in the beginning God created the heaven and the earth."

Bibliography

Atkins, P. W. 1995. *The Periodic Kingdom*. New York: Basic Books.

Ball, P. 2002. *The Ingredients: A Guided Tour of the Elements*. New York: Oxford University Press.

Bouma, J. 1989. An application-oriented periodic table of the elements. *Journal of Chemical Education*. 66: 741.

Brady, J. E. and G. E. Humiston. 1986. *General Chemistry: Principles and Structure*. Hoboken, NJ: John Wiley & Sons.

Brown, T. E., H. E. LeMay, and B. E. Bursten. 2005. *Chemistry: The Central Science*, 10th ed. Upper Saddle River, NJ: Prentice Hall.

Carithers, B and P. Grannis. 1995. Discovery of the Top Quark. *Beam Line* (SLAC). 25 (3): 4-16.

Chandrasekhar, S. 2003. *Newton's Principia for the Common Reader*. New York: Oxford University Press.

Dalton, J. 1842. *A New System of Chemical Philosophy*, 2nd ed. London: John Weale.

Gish, D. 1978. Thermodynamics and the Origin of Life (Part II). *Acts & Facts*. 7 (4).

Griffiths, D. J. 2008. *Introduction to Elementary Particles*, 2nd ed. Weinheim, Germany: Wiley–VCH.

Lindsay, R. B. 1959. Entropy Consumption and Values in Physical Science. *American Scientist*. 47: 378.

Mayer, R. 1841. *Remarks on the Forces of Nature*. Cited in Lehninger, A. L. 1971. *Bioenergetics: The Molecular Basis of Biological Energy Transformations*, 2nd ed. Reading, MA: Addison-Wesley.

Mazurs, E. G. 1974. *Graphical Representations of the Periodic System During One Hundred Years*. Tuscaloosa, AL: The University of Alabama Press.

Carnot, S. 2005. *Reflections on the Motive Power of Fire: And Other Papers on the Second Law of Thermodynamics*. Mineola, NY: Dover Publications.

Morris, H. 1978. Thermodynamics and the Origin of Life (Part I). *Acts & Facts*. 7 (3).

Newlands, J. A. R. 1865. On the Law of Octaves. *Chemical News*. 12: 83.

Pullman, B. 1998. *The Atom in the History of Human Thought*. Reisinger, A., trans. New York: Oxford University Press.

Riordan, M. 1987. *The Hunting of the Quark: A True Story of Modern Physics*. New York: Simon & Schuster.

Scerri, E. R. 2007. *The Periodic Table: Its Story and Its Significance*. New York: Oxford University Press.

IMAGE CREDITS

NASA: 15, 42

United States Department of Energy: 21

Susan Windsor: 18, 20, 22, 30, 37, 38, 39, 41

FOR MORE INFORMATION

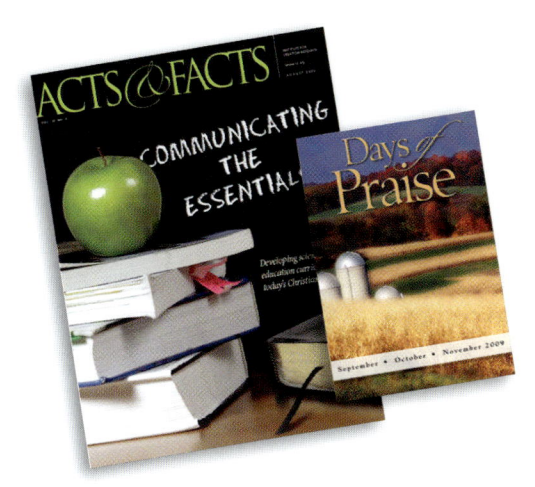

Sign up for ICR's FREE publications!

Our monthly *Acts & Facts* magazine offers fascinating articles and current information on creation, evolution, and more. Our quarterly *Days of Praise* booklet provides daily devotionals—real biblical "meat"—to strengthen and encourage the Christian witness.

To subscribe, call 800.337.0375 or mail your address information to the address below. Or sign up online at www.icr.org.

Visit ICR online

ICR.org offers a wealth of resources and information on scientific creationism and biblical worldview issues.

✓ Read our daily news postings on today's hottest science topics

✓ Explore the Evidence for Creation

✓ Investigate our graduate and professional education programs

✓ Dive into our archive of 40 years of scientific articles

✓ Listen to current and past radio programs

✓ Order creation science materials online

✓ And more!

For more information, contact:

INSTITUTE for CREATION RESEARCH

P. O. Box 59029
Dallas, TX 75229
800.337.0375

SCIENCE EDUCATION ESSENTIALS

How do I explain the differences between biblical creation and evolution?
What evidence for the origin of life should my students know?
Where do I go for trustworthy information on science research and education?

For 40 years, the Institute for Creation Research has equipped teachers with evidence of the accuracy and authority of Scripture. In keeping with this mission, ICR presents Science Education Essentials, a series of science teaching supplements that exemplifies what ICR does best—providing solid answers for the tough questions teachers face about science and origins.

This series promotes a biblical worldview by presenting conceptual knowledge and comprehension of the science that supports creation. The supplements help teachers approach the content and Bible with ease and with the authority needed to help their students build a defense for Genesis 1-11.

Each teaching supplement includes a content book and a CD-ROM packed with K-12 reproducible classroom activities and PowerPoint presentations. Science Education Essentials are designed to work within your school's existing science curriculum, with an uncompromising foundation of creation-based science instruction.

Demand the Evidence. Get it @ ICR.

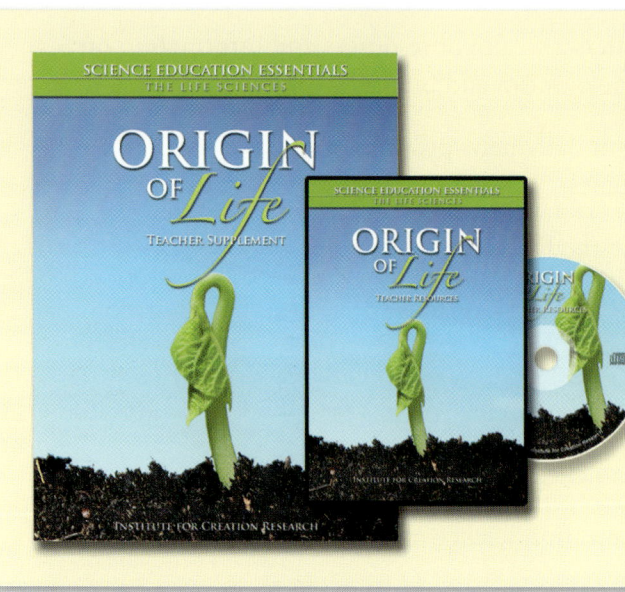

Origin of Life

How did life get started on earth? Many scientists believe that life began from natural processes, but the Bible presents an alternate explanation.

Origin of Life, the first of the series, answers basic life questions such as:

- What is the origin of life?
- What are the physical and biblical definitions of life?
- What are the physical requirements for life?
- Can life exist elsewhere in the solar system?

It gives scientific explanations for the chemical basis for life from a biblical worldview and discusses the efforts to create life in the laboratory. Most importantly, it offers scientific evidence proving that the creation of life "requires an act of God."

Available for **$24.95** (plus shipping and handling)

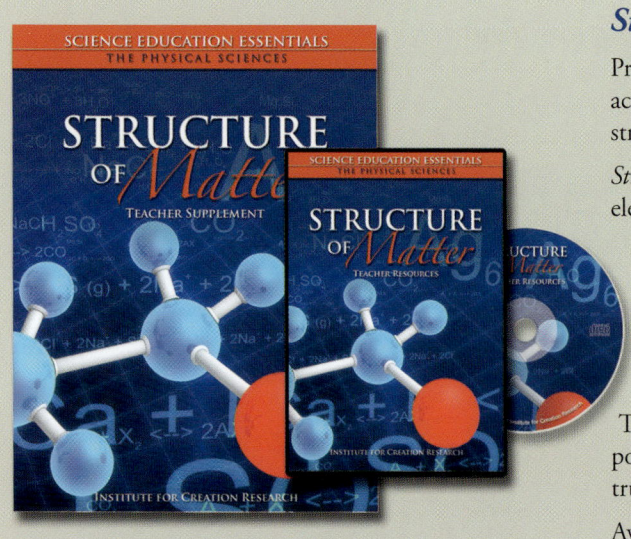

Structure of Matter

Predictions in science are based on knowledge of observable events. The accuracy with which science can make predictions points to the order and structure God established within His created universe.

Structure of Matter, the second of the series, explores structural forces and elements of nature such as:

- The First and Second Laws of Thermodynamics
- The structure of the atom
- The periodic table
- Properties of matter
- And more

The order and design of the universe point to a Creator of omnipotent power and omniscient strength. Truly, the structure of matter upholds the truth that "in the beginning God created the heaven and the earth."

Available for **$24.95** (plus shipping and handling)

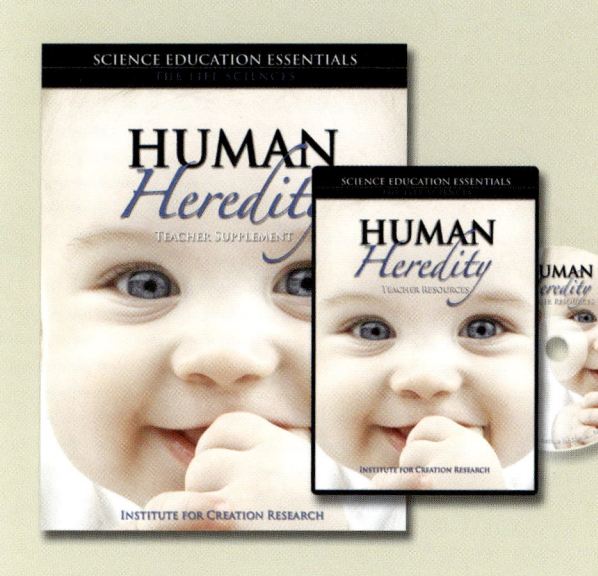

Human Heredity

Genes provide most of the information that determines physical appearance and even influences certain behaviors. In spite of the differences among humans, their genomes are still 99.9% identical. Did everyone come from two people?

Human Heredity, the third of the series, examines such topics as:

- Our inheritance from our parents
- Dominant and recessive traits
- Human descent from Adam and Eve
- Polygenic inheritance
- And more

The study of genetics has expanded our understanding of human inheritance, leading to the inevitable conclusion that all humans came from the first created man and woman. "And Adam called his wife's name Eve; because she was the mother of all living" (Genesis 3:20).

Available for **$24.95** (plus shipping and handling)

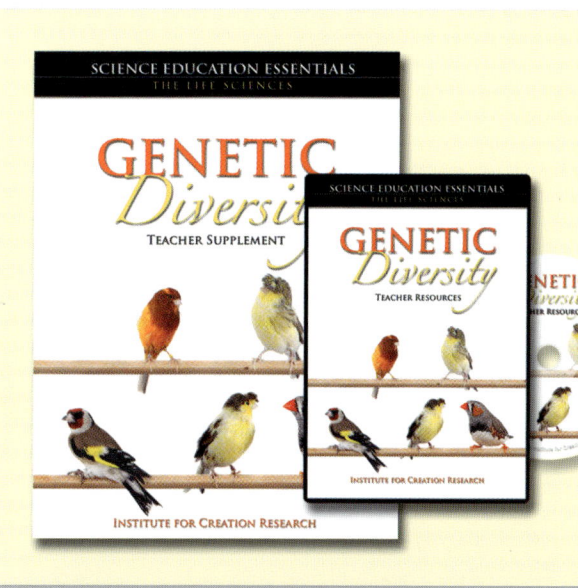

Genetic Diversity

God created an incredible variety of incredible creatures—and it seems He created us in His image to enjoy that variety. What is the science behind this wonderful diversity?

Genetic Diversity, the fourth in the series, takes an in-depth look at:

- The classification of living things
- Differences among species and within kinds
- Diversity and the mosaic concept
- And more

When it comes to the kinds of life observable today, the overwhelming scientific evidence strongly supports the thinking of creation scientists that life can only come from the eternal Creator.

Available for **$24.95** (plus shipping and handling)

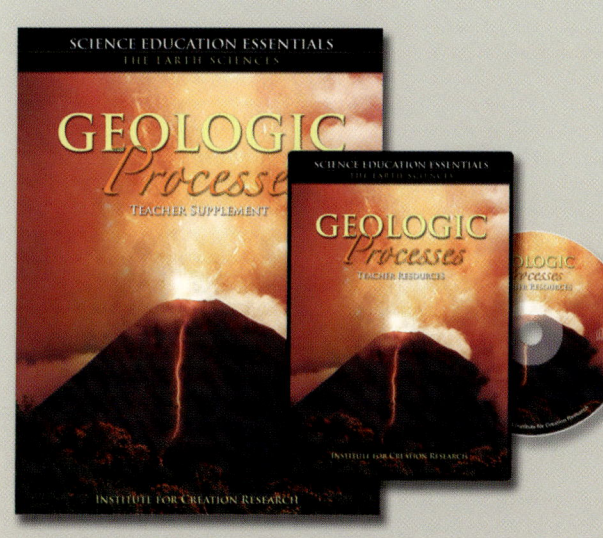

Geologic Processes

What geologic processes shaped our earth? Is evolution right, that it developed gradually over millions of years? Or does the geologic record demonstrate something else?

Geologic Processes, the fifth in the series, studies earth's history to answer the questions:

- Did the earth start as a cosmic collection of star dust?
- What processes shape the earth today?
- What types of rocks are found on earth?
- What is the geologic evidence for a worldwide flood?
- And more

The world is a museum of past processes that operated on a much greater scale, proceeded at a more rapid rate, and acted with more intensity than those acting today. The best explanation for earth's history is the biblical record of creation and the great Flood.

Available for **$24.95** (plus shipping and handling)